NANO
IN A
NUTSHELL

AuthorHouse™
1663 Liberty Drive
Bloomington, IN 47403
www.authorhouse.com
Phone: 1 (800) 839-8640

Published by AuthorHouse 12/02/2015

ISBN: 978-1-5049-6431-9 (sc)
ISBN: 978-1-5049-6430-2 (e)

Library of Congress Control Number: 2015919533

Print information available on the last page.

Any people depicted in stock imagery provided by Thinkstock are models,
and such images are being used for illustrative purposes only.
Certain stock imagery © Thinkstock.

This book is printed on acid-free paper.

NANO
IN A
NUTSHELL

A Young Student's Guide to Nanotechnology
Tools, Applications & Challenges for the 21st
Century and Beyond

ANDREA J HARMER

Hi, it's me again, Nan, and of course my sassy sidekick, Buckydog! I'm in high school now, studying nanotechnology, and Buckydog is a lot older too. It's been a long time since we chatted together, welcome back!

Because you have chosen to read this book, you already have a curious and open mind. Way to go, because the science and technologies I'm about to tell you about are truly amazing and you will need to keep your mind open simply to comprehend them! What is even more remarkable is that these new technology applications, enabled by science, are actually true! These technologies are made possible by studying, applying, and controlling matter on the *nanoscale*, known as *nanotechnology*. I'm here to explain them to you in a nutshell, or rather as simply as I am able. Let's get started...

As we know from our studies before, **nanotechnology is the control of matter and structures at the nanometer scale** and **one nanometer is one billionth of a meter.** One meter = 3.28 feet, so one nanometer is one billionth of approximately 3 feet! For example, your hair is about 50,000 – 100,000 nanometers thick. The **diameter of one single atom, depending upon which element it is from, is approximately between 1/10th to 2/3 nm.**

(Scientists use radius, or 1/2 diameter.)

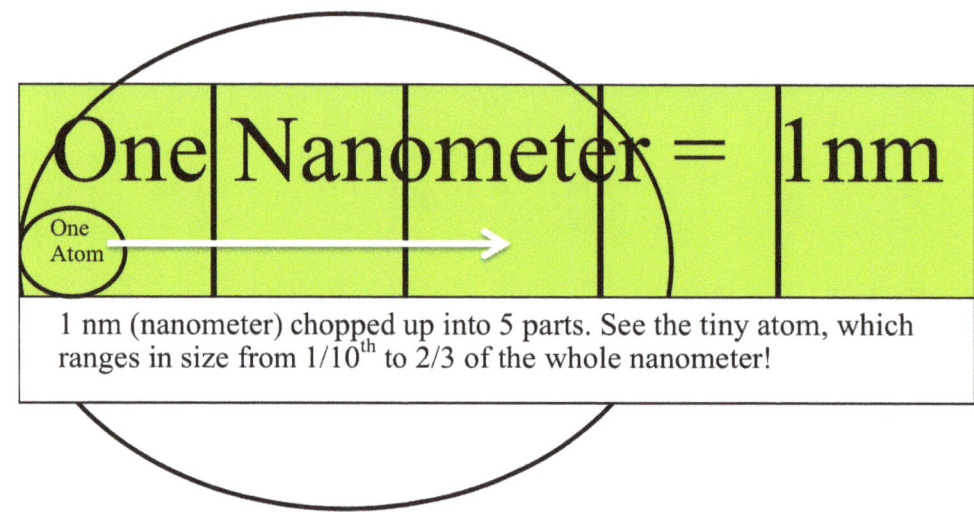

One Nanometer = 1 nm

One Atom

1 nm (nanometer) chopped up into 5 parts. See the tiny atom, which ranges in size from 1/10th to 2/3 of the whole nanometer!

Wow, my hair is 50 thousand nanometers thick!! That's amazing!! Man, how could anybody control anything *that* small ?

During the past 10-15 years, powerful microscopes were invented using the same technology as the famous Hubble Microscope in outer space. These microscopes now allow us to see atoms and molecules that we could never see before, and guess what! The real molecules look just like the models that scientists developed using X-rays long before they could even see the molecules for real. That is way cool. Check it out ...

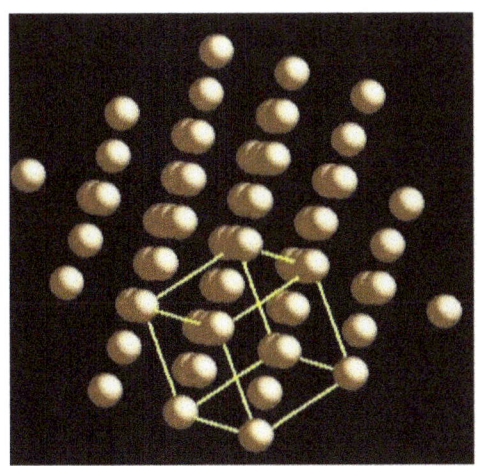

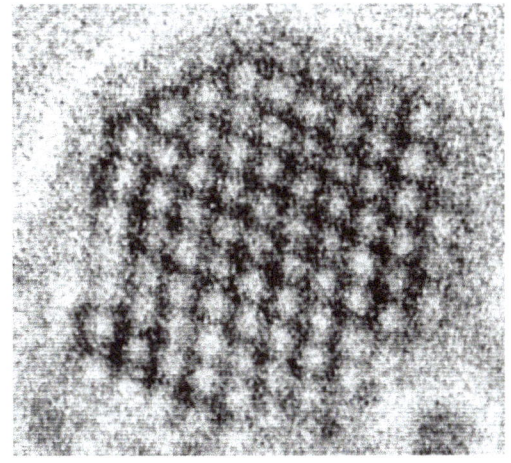

On the bottom left of page 3 is a picture of the molecular model of gold that scientists modeled before they could truly see what gold atoms looked like, and on the right is a real picture of that same gold *molecule* taken with a *transmission electron microscope*, also known as a TEM.

A TEM uses electrons to look through a sample molecule in the microscope, so the sample has to be very thin without breaking apart. Making samples that thin is a real challenge for scientists. The person with the thinnest samples gets to see the most in the microscope! In addition, the person who studies all the tricky, *abnormal* parts of the molecule in the microscope, gets to learn the most too! A lot of science is discovering patterns, whether normal or abnormal. On the next page is a picture of a state-of the-art TEM.

A powerful *scanning electron microscope* (SEM) and an *atomic force microscope* (AFM) are also able to see on the nanoscale. The SEM bounces electrons off the sample to make an image we can see. The AFM scans the surface of the sample so that we can see how smooth or bumpy it is.

Transmission Electron Microscope, also known as The ARM, short for Atomic Resolution Microscope, *courtesy of Lehigh University*

Anyway, where are we going with all this? What can we do with all these atoms and molecules now that we can see them and are just learning how to control them?

Yeah, Let's get to the neat stuff. What's the big fuss about nanotechnology? What can I do with this? Nano biscuits don't sound all that appealing! When it comes to food, I like bulk! *Bulk or macroscale* is what we call matter if it is larger than the nano or the microscale.

Well, the *nanoscale* deals with particles of matter that are between 1 and 100 nanometers or, as scientists say, between 1 –100 nm in size. The *microscale* is a little larger, between 101 nanometers to 100 micrometers. A single micrometer is called a micron, written as 1 µm.

The *big fuss about nanotechnology* is that sometimes atoms and molecules on the nanoscale behave very differently from the way that scientists expect

them to behave, based on scientists' current knowledge. This means that there are completely new and different ways of using nanoscale matter for making new technologies. This is the foundation of the new scientific knowledge that is now emerging for creating things using the nanoscale. This is nanotechnology!

What this means for you is that you could be the next one to discover something completely new, based on the unknown and unusual behavior of atoms and molecules on the nanoscale.

That is ultra cool, don't you think?

OK, OK. Enough about science already!!! What's this stuff good for anyway?

Well, Mr. Buckydog, how about helping people who are struggling with cancer, or trying to clean up water and soil that has been polluted by years of misuse?

How about inventing solar cells that could capture energy so efficiently that we could use sunlight to heat our houses instead of Earth-polluting fossil fuels, like oil and gas? How about making fuel cells for batteries that could replace gasoline in cars? How about making a cell phone battery that would only have to be charged once a year?

And get this, how about putting something in your socks that would keep your feet from getting smelly!!

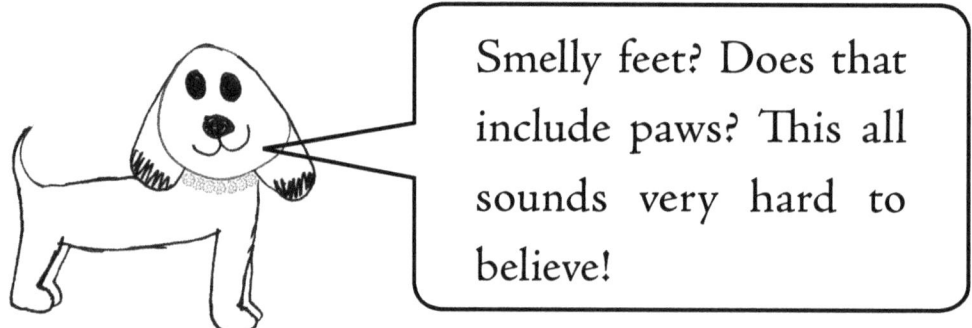

Smelly feet? Does that include paws? This all sounds very hard to believe!

Well, it's true. Scientists right now are working on each of these applications (and more!) using nanotechnology. In a minute, I will introduce you to one of them, along with her research. This might help you understand how scientists are applying nanoscale science to help make our lives better.

Before we begin, one of the interesting ideas you should understand is that nanotechnology seems to be affecting many *different* fields of science, such as medicine, electronics, the environment, and even the materials we use everyday, like the clothing we wear, or the cell phone screens we use constantly.

Hmm, I wonder what they will think of next!

When scientists get together and discuss their research, they usually figure out among themselves which field of study interests each scientist the most. In fact, the most interesting ideas typically form when scientists from different fields or different *disciplines* of science get together and discuss how each of them could apply their own unique knowledge to a shared problem, such as curing cancer, global warming, or finding a renewable energy source.

This is known as *interdisciplinary* research because several scientists come together from different scientific *disciplines* and *collaborate* together to solve problems. This is a very common and useful practice in science, technology and engineering.

Do you mean that if John, who understands radio waves, talks to Steven, who operates on cancer patients, talks to Richard who discovered nanoparticles, that we might be able to treat cancer using nanoparticles through radio waves?

Exactly, Bucky!! Now you get the idea! In fact that is exactly what happened when John, Steven, and Richard did talk!!

Mixing it up can be very beneficial in science. I bet you never thought of it that way before!!

Of course, you can't just go mixing chemicals before you know what kind of reaction you might get, that could be dangerous and bad for the environment!! However, once you know the foundations of science, your own imagination is the only limit that you may know.

Cool! My imagination is one of my best assets!!

Ok, now that we understand that using your imagination in science is a great idea, close your eyes and imagine how you might use the smallest particle of matter that we know exists today.

Would you use it to repair your grandmother's failing eyesight? Would you use it to keep germs off your Mom's kitchen counter?

Would you use it to stop your knee from bleeding when you fall off your bike?

Would you use it to fill in the cavities in your Dad's teeth?

What would you use nanoparticles for??? Think about it…..

Nanotechnologists around the world are working on some of these applications and more, but the really neat part is that it is all happening right now! It's like, if nanotechnology were a baseball game, we'd still be in the first inning! And each of you, who love studying science, could be the next big hitter! Or, if nanotechnology were a flower, we would just be seeing the bud. This means that your ideas about how to use atoms and molecules at the nanoscale could be the next big, blooming idea! Not everything in science has been discovered. In fact, there is much more to be discovered today than ever!

Wow, I bet Benjamin Franklin would like to be alive today! He loved using science to invent new technologies! Just imagine what he could do with a nano-sized lightning rod!

Now, let's look at some specific scientific *disciplines* and see how nanotechnology is being studied and why. First, let's take a look at how nanotechnology is being used in biology.

Bio means life, and "ology" means "the study of. " Put them together and what do you have?

Biology = the study of life.

When it comes to nanoscale biology, some scientists call it "nano-bio," for short.

One of the most interesting applications in "nano-bio," is how gold *nanoshells* are being used (or *applied*) to detect and treat cancer cells more effectively. This is really good news for the people who have cancer! Typically, our bodies don't react badly to the *metallic* element, gold, so doctors and scientists use gold nanoshells as a vehicle, like a very very tiny car, to deliver toxic drugs to kill cancer cells. The nanoshells are coated with iron oxide (rust is an example of an iron oxide) to keep them from destroying good cells along their ride. Once the gold nanoshells reach the bad cancer site, scientists and doctors use light waves to release the toxic drugs in the nanoshells to kill the cancer cells.

Get this, the doctor and researcher who *discovered* gold nanoshells and is using them to detect and target cancer cells is a woman named *Naomi Halas*. Doctor Halas and her all-female team of other doctors live and work in Texas at a place called Rice University. There, they also study how the beautiful, colorful, *optical* properties of gold metallic nanostructures can be engineered to capture and harvest sunlight (called *solar*) for new energy.

They also study how metallic nanostructures can sense toxic chemicals to keep us safe at home and in the environment. In ancient times, people used metallic nanostructures to create beautiful windows and jewelry, without even knowing what the actual structures looked like. Today, using TEMs and SEMs, Naomi Halas and her team are studying nanostructures to make state-of-the-art computer chips that will make our electronic devices hold even more information and work even faster!

You may remember from our last discussion that the word *optical* means related to light. Well, *optics* is the science that describes the behavior and properties of light and the way light interacts with matter on the *bulk* scale, or the scale that we can see with our *naked eye*.

When it comes to the nanoscale, that same study of *optics* is called *nanophotonics,* or as some people like to call it, nano-opto or opto-nano for short.

In other words, *nanophotonics* is the study of the way that light behaves on the nanoscale, which is different from how it behaves on the *bulk* or larger scale.

Think of it like this, *photonics* is almost like the word photography, which means recording with light. The way that we take a picture is by using the energy in light to record an image. The energy is called a *photon*. A photon is the particle that is responsible for producing the light's energy, which is also called electromagnetic energy. Well, to sum it up, *photonics* is basically the science of generating, controlling, and detecting photons. So what do you think they call it when they do this on the nanoscale? Come on, you know! Yep, that's right, nanophotonics!! Cool! Good job!

Way to go, Smarty pants! A++

Dr. Halas uses the brand new field of nanophotonics to generate and manipulate light at nanoscale dimensions. Awesome! So what else is happening in nanotechnology?

In the electronics field, which is where all of our cool hand devices, like cell phones and tablets and smart watches are created, scientists have traditionally used a "top down" approach to create these items. From an engineering viewpoint, a "top down" approach is simply trying to make the device smaller and smaller and smaller and smaller. The problem is that now, after nearly 60 years of using a "top down" approach, scientists have just about reached their limit of creating technology this way.

Because scientists have nearly reached their limit with "top down" electronics, also called *microelectronics*, they are now looking toward the "bottom up" *nanoelectronics* approach. Nanoelectronics is the study of how to control atoms and molecules on the nanoscale to make even smaller and faster electronics.

As we mentioned before, nanoelectronics uses the "bottoms- up" approach, attempting to assemble atoms and molecules to create even smaller chips for tinier electronic devices. The biggest challenge with creating even smaller electronic chips is overcoming something called the *"diffraction wavelength."*

Yay, great idea, I need a smart watch small enough to fit on my leg so I can get home in time for dinner, my favorite part of the day, and to keep track of my daily walks that keep me fit! Maybe I need a nano chip!

Hang on a minute, … I think I'll have a potato chip instead, hee, hee!

Diffraction is the bending of light around obstacles and through slits. Because light travels in waves, when it hits obstacles, it has to bend and that's when it is interfered with. This interference is called *diffraction*. Diffraction can happen with any thing that travels in waves, such as light, sound, X- rays, radio waves and even the ocean waves! Anytime these waves are interfered with, there is diffraction occurring.

At the atomic level, all physical objects have wave-like properties and when they are interfered with, there is diffraction. The study of diffraction, and other principles at the nanoscale is called *Quantum mechanics*. What's cool about quantum mechanics is that it studies and explains the way that matter behaves and interacts at the nanoscale, which is very different from the bulk scale we see with our eyes. This is the true mystery of the emerging study of nanoscale science and the way we apply it in nanotechnologies!

Maybe you will be the one to be able to figure out how to control or manipulate diffraction for smaller, faster electronics using quantum mechanics, also known as quantum physics or quantum theory. Keep studying your physics, chemistry and math. If you do, you will become a science superhero!!

Superhero!!

I like the sound of that!

Go for it!! You will be glad you did!!

"Be a scientist,
save the world!"

-- Dr. Richard
Smalley

Here is a **Glossary** to help you in your studies!

Aberration Corrector – a new fitting on a powerful electron microscope that measures 'over-focus' in the microscope and adjusts the electron beam accordingly. The resulting beam is only a tenth of a nanometer wide.

Atom – the smallest particle of an element that can exist.

Atomic Force Microscopye – a powerful microscope that scans the surface of a sample so that we can see how smooth or bumpy it is, even on the nanoscale.

Buckyball –a soccer ball shaped molecule made up of 60 carbon atoms, discovered in 1985, also known as the C60 buckyball or C60 molecule, the building block of nantechnology.

Bulk or macroscale – structures that are larger than the nano or microscale.

Cheeky Monkey – a British term meaning a mischievous one.

Diffraction – the bending of light around obstacles and slits, light travels in waves, this wave interference because of an obstacle is called diffraction, can also happen with anything that travels in waves, like sound, x-rays, radio waves.

Electron –one of three parts found in an atom, the other two parts are called a neutron (like Jimmy Neutron!) and a proton, electrons have a negative charge.

Electron Microscope – a microscope that magnifies things by using electrons to form an image instead of light (the microscope that uses light is called a light optical microscope).

Element – a pure chemical substance that contains only one kind of atom (see http://insideout.rigb.org/ri/elements/inside/index.html for more information about elements) **Fuel cell** - a device that converts the chemical energy from a *fuel* into electricity through a chemical reaction.

Interdisciplinary Research – research that crosses several boundaries of fields of study at once, bringing in different perspectives and multiple, collaborative expertise.

Metallic –relating to metals.

Microelectronics – the study of how to control atoms and molecules on the microscale to build small and efficient electronics.

Micrograph – an image or photograph that is seen or captured through a microscope.

Micrometer – a unit of measure, also called a micron, ex. Human hair is typically 80-100 micrometers.

Microscale– the scale of measurement between 101 and 100 micrometers.

Molecule – a unit of matter, a structure of assembled atoms.

Naked Eye – an eye that looks at things without the help of a microscope

Nano – means dwarf.

Nanoelectronics – the study of how to control atoms and molecules on the nanoscale to make even smaller and faster electronics, such as cell phones.

Nanometer – a spatial measurement that is equal to one billionth of a meter.

Nanoparticle – a particle of matter that is between 1 and 100 nanometers in size.

Nanophotonics – the science of generating, controlling, and detecting photons on the nanoscale.

Nanoshells – a type of spherical nanoparticle consisting of a core, which is covered by a thin metallic shell.

Nanoscale – generally refers to structures with a length scale of 1-100 nanometers.

Nanoscience – scientific knowledge based on the smallest scale man is now able to see and change.

Nanotechnology – applying nanoscience principles to making things, scientists say: nanotechnology is the control of matter and structures at the nanometer scale.

Nanotube – a nanoform discovered in 1991, the fourth form of carbon known to man.

Naomi Halas – first one, with her team of scientists, to discover nanoshells in 2003, "We said, 'Gee, what could it be good for?'" Halas told CNN (Wikipedia, 2015). Now gold nanoshells are used for cancer detection and targeted drug delivery.

Optical – relating to vision, optics is the science of light.

Photon - an elementary particle, the force carrier for the electromagnetic force or energy.

Photonics - the science of generating, controlling, and detecting photons.

Quantum Mechanics – study of diffraction and other principles at the nanoscale, explains the way that matter behaves and interacts at the nanoscale.

Acknowledgements:

Many thanks again to my husband, Martin P. Harmer, for his scientific expertise.

 Many thanks to Lauren Stetler for always believing in Nan and Buckydog.

Many thanks to my family, for their continued support and patience.

Credits:

Nan and Buckydog created by Andrea Harmer.

Nanometer graphic (p2) illustrated by Andrea Harmer.

All images of Buckydog illustrated by Laura Pressler.

All images of Nan (grown up) illustrated by Lauren Stetler.

References:

Cover graphic created as part of NASA sponsored Nanotechnology

Exhibit for the DaVinci Discovery Center in Allentown, PA.

Molecular model of gold, A. Maeland and T.B. Flanagan, Can. J. Phys.,1964, 42, 2364. Mark Winter, University of Sheffield, UK.

Photograph of gold molecule and TEM ARM microscope, courtesy of Lehigh University. (C. Kiely and Mary Anne Lynch).

Smalley quote retrieved September 25th, 2015 from http://cnst.rice.edu/founder/

Did you know?

Your fingernail grows one nanometer in one second!! Check out more facts about the Size of the Nanoscale at http://www.nano.gov/nanotech-101/what/nano-size

Now it's Your Turn!

On these pages, draw what you would make using nanoparticles.

Use Your Imagination! Be a Scientist, Change the World!